Victoria and Albert Museum

Keeping Warm

The Arts and Living

London: Her Majesty's Stationery Office

Acknowledgements

Alastair Laing started me off with some
interesting information. Jeremy Lever kept me
going with constant explanations of technical
matters. I am grateful to them and to the staff
of the Victoria and Albert Museum for their
generous assistance.

With the following exceptions all the objects
illustrated are from the Victoria and Albert
Museum. Figs. 3, 4, 5 and 21 Science Museum.

Design by HMSO Graphic Design

Cover. Bellows of floral marquetry, the reverse
with embossed silver. Made 1675 for the Duke
of Lauderdale, Ham House, Surrey.

ISBN 0 11 290293 6

Contents

1 Some like it Cool

Man is compelled by the elements to create an artificial climate indoors, and by human nature to debate forever its merits and faults. Warmth is a nebulous medium between the positively unendurable extremes of heat and cold. It is differently defined by each individual according to his age, size, smoking and eating habits, general health and mobility, according to the standards and customs of the time and to a multitude of other factors. Modern scientists have analysed the properties of heat and measured precisely the effects of the slightest changes in temperature upon every conceivable material from epidermal tissues to net curtains. Yet they cannot provide anything more precise than a general estimate of our present 'comfort zones' to guide heating engineers. How then is the historian to assess the comfort of previous generations? Not with a thermometer already raised to a high degree.

Houses of the past were doubtlessly draughtier than ours and their temperatures sometimes lower. But their occupants, differently dressed and following a totally different pattern of life were not necessarily uncomfortable. On the contrary, since Roman times there have been more complaints of excessive artificial heat than of natural cold.

A sermon on the virtues of coolness and the evils of heat may not only be appropriate to our present energy crisis, it may prove, as many sermons do, that the evil has always been with us.

'Live they not against Nature that long for a rose in winter' warned Seneca as his fellow Romans luxuriated in the perpetual summer produced by their subterranean furnaces or hypocausts. Technology has given us much more control over nature. Yet we still cling to the seasons. We rejected Le

4

Corbusier's proposal to house us, all over the world, in a uniform climate all year round. We prefer to revel in Spring fever which was regarded by Catherine Beecher in 1869 as a moral punishment visited upon Americans for breathing the hot vitiated air of their stoves rather than the cool fresh air ordained for their lungs by God.

Egyptians, believing coolness brought the soul closer to heaven, aired their heads on hard pillows. From cold, Aristotle observed, 'the flesh gains firmness, and solidity.' Vigour, strength, hardiness, courage – the cardinal attributes of manliness, prized by kings, knights and peasants – all required cold. Heat, oppressive, softening, emasculating, was for the old and infirm, for tender women and flowers. Cold, because it toughened the constitution against illness, was considered healthier than debilitating heat. Draughts, although feared by the weak, did not disturb the hardy. North winds that 'purgeth vapours' were encouraged by Andrew Boorde in 1542 when people were careless in 'emptying chamber pots and pissing in the chimney', when spitting on the floor was commonplace, and plague rife.

Numbing cold is the plight of the very poor. Heat displays the material achievements of the middle classes. But coolness was favoured by men of rank to whom the preservation of health and strength was of vital importance. Chinese emperors distinguished themselves by keeping lower indoor temperatures and larger stores of furs. The Sun King was certain not to melt at Versailles. George III, to strengthen his constitution, deprived all but four apartments in Buckingham Palace of 'carpets and other means of great warmth.' Queen Victoria demanded international respect for her hatred of heat.

Today, coolness has lost its value and draughts have been excluded. But heat is no less worrisome. This once dreaded robber of health, having been accepted as a household companion, is currently suspect of plundering our natural resources.

2 Warmth without Fire

Food is the ultimate source of body heat. Nature purposefully provides calorific winter fuels – fattened pigs and pounds of potatoes which were consumed in quantities before the invention of artificial summer. Stimulating exercise also raises the temperature. Where short, bitter days curtail outdoor activity, strong drink – schnaps, vodka, whiskey – is the common anti-freeze and anti-depressant. Mild wines suffice for Mediterraneans.

Clothing is a preserver not a producer of warmth. Its function, though often thwarted by fashion, is never totally defeated. Medieval men wearing short trousers to display their virility, had trunk hose and robes for warmth as 1960s mini-skirted girls had tights and boots. Sedentary members of society – women, priests, scholars, etc – required the extra protection of long garments *[pls.1,2]*. The occasional dropping of drawers and necklines did not seriously diminish the insulation provided by numerous layers of clothing – chemises, doublets, petticoats, skirts and dresses, waistcoats, jackets, shawls, night (meaning indoor) gowns, coifs and caps, made of substantial fabrics quilted, lined with sheepskin or sable according to one's means.

Flannel puttees, one or more pairs of heavy stockings, perhaps with gaiters or leather insteps, protected the feet, the coldest part of the body, from draughty floors. Fur boots were commonplace in Russia. For Cinderella *vair* (fur), misconstrued as *vere* (glass), was a luxury. On getting out of bed, when sitting to eat, read or write the feet would have quickly surrendered their hard gained warmth to the floor were it not for the intervention of a ledge, rug, cushion or footstool. A footbearer cherished the feet of a Medieval Welsh nobleman prince in his bosom. A bearskin sack prevented Liselotte's

6

toes from freezing as the wine did at Versailles in 1707. A dog or cat served just as well. Foot warming braziers, metal containers filled with charcoal embers and covered with wood, were in continuous use from Roman to Edwardian times *[pl.3]*. Hot water bottles were safer, especially in bed. Shoes fitted with hollow wooden soles containing 'mars well rubified' (red hot metal) enabled Philip II of Portugal (III of Spain) to go hot footed indoors and out.

Busy hands needed less comforting than idle feet. During a long sermon or ceremony, however, they did well to have the warmth of an ember box suspended on gimbals, a contrivance to keep it always in a horizontal position, within an apple of copper or brass *[pl.3]*.

A house is essential protection against cold, and the ultimate container of both natural and artificial warmth. Nevertheless, architectural design, like *haute couture*, is often inattentive to this function. Insulation has long been regarded as the business more of the occupant than the builder or architect. The wise man was counselled by Alberti to build rather for summer than winter, and 'arm . . . against the cold by making all close and keeping good fires.' This tacit division of labour worked well enough as long as supplies of fuel were equal to demands of comfort. Now that they are not, architecture may have to become the devoted, even the legal husband of warmth.

Man's determination to make close in winter has never yet been defeated by his need for defence or desire for show. Interiors in the past were as heavily layered as their occupants. Walls have been plastered since antiquity. Tapestries and hangings of heavy fabrics were an essential part of the portable warmth that followed the nomadic people of the Middle Ages from castle to castle, and remained in use in great houses until the end of the seventeenth century. Increasing stability after the thirteenth century permitted the permanent insulation of wainscot backed with wool and charcoal, of tooled leather, cut velvets, arras and damask fixed on stretchers. Moreover, these warm textures and rich dark

7

colours gave an appearance of snugness to large draughty apartments; but exacerbated the fug of smaller, better sealed, heated and furnished nineteenth-century rooms.

Window glazing, introduced by richer Romans, remained too expensive for ordinary domestic use until the late Middle Ages. Alabaster, horn, oyster shell, linen, paper or silk in the Orient provided protection from the elements without the total loss of light suffered by the poor with nothing but wooden shutters. An inner window frame of oiled paper or muslin was the precursor and common alternative to double glazing which was known in the mid-eighteenth century to the French rationalist, Abbé Laugier, probably from his residence in Germany. Glazed conservatories contributed much to the warmth of nineteenth-century houses. Windows were not only shuttered at night, but curtained as well, the best with rich weighty fabrics, lined, even interlined, and finished above with valences so generously draped as to form another curtain. In 1601, at Hardwick Hall 'all glass and no wall' Bess, Countess of Shrewsbury made her bed chamber snug with two curtains of warm red cloth and 'three coverlets to hang before a windowe.' Wads of cloth or plain paper and paste were common home remedies for ill-fitting windows before self-adhesive plastic strips came into the shops.

Doors, like windows, were traditionally curtained with *portières*, and sprung, from the eighteenth century onwards, to snap closed. Double doors succeeded inner porches *[fig.1]* in providing permanent draught free access. Skirtings that rose and fell as doorknobs turned were an improvement upon the homely sausages that guarded the gap.

Floors, until the end of the sixteenth century, were covered only by rush mats. Woven or knotted carpets, although too precious to be continuously walked over, were occasionally removed from the tables and beds where they normally lay for temporary use as foot rests alongside beds, dining tables and hearths. A floor cloth covered the dais before the hearth where Henry VIII dined. In her bed chamber, Bess of Hardwick had 'two foote Carpets of turkie worke' and eight

Figure 1. Draught-proofing inner porch from Sizergh Castle, Westmorland, 1382.

'fledges' or down quilts around her well enclosed bed. Although carpeting increased in the seventeenth century, it remained a luxury until the introduction of machine manufacture at Kidderminster in 1735. Thereafter it spread from reception rooms to bedrooms, into passages and on to the stairs.

Electric carpets seemed to have had limited currency, probably among those jaded by plush. Underfloor heating provides gentle background warmth at punishing cost. Thus, like Roman hypocausts, its use is restricted to the richest homes and even there to a few rooms.

Rooms were not only lined, but also partitioned for warmth. Permanent timber screens were built across draughty entrances to medieval halls and chambers. Travices – partitions of heavy cloth or tapestry – were temporarily suspended to cut cold corners [pls.1, 2]. Fires were also sheltered by portable screens. The majority for daily use were made of modest wicker or wood. The best, for public view, might be as tall, broad and ostentatious as the mantelpiece. Large cabinets and high backed settles were brought out from the walls in winter to make cosy chimney corners. To Thomas Hardy, the settle was to the old fashioned fireplace what 'the east belt is to the exposed country estate, or the north wall to the garden. Inside is Paradise. Not a symptom of a draught . . . the sitters' backs are as warm as their faces . . .'

Cosiness, however, was neither found nor expected everywhere. Great halls, saloons, long galleries and such were expressly designed for public activity, parade and reception which generated its own heat. Bess of Hardwick would never have attempted to spend a cosy evening in her long gallery in 1597 as the 6th Duke of Devonshire did in 1844. Understandably he failed although 'surrounded by screens and sheltered by red baize curtains'; rightly he sought refuge in the bed chamber where Bess lived. Likewise, today's occupants of ancestral mansions must either adapt their pattern of life to the house, or equip it with central heating.

Before central heating and air conditioning made our indoor climate uniform, rooms, furnishings, even houses were, like clothing, changed winter and summer. Chinese emperors of the third century B.C. spent their lives rotating clockwise with the sun through a series of seasonal apartments. Egyptians rose to their house tops in summer to catch the breezes and escape insects, and descended in winter to be near the kitchen fire and out of the wind. Greeks and Romans moved horizontally. The winter wing of Pliny's villa was sheltered from the cold, orientated towards the sun and bow-windowed to capture its full warmth. According to Vitruvius winter apartments, blackened by the smoke of fires, needed little decoration, but they did require, in Greece, extra porous paving to absorb ritually spit and spilled wine. Many European houses, although thoroughly furnished with fireplaces, were not thoroughly occupied in winter. Front and back parlours were, and still are seasonally used. Recalling past practice, Alexander Pike's energy saving Antarkic House of the future has an inner winter core wrapped in a light summer cover.

French fashion, imitated throughout Europe and America, demanded complete change of furnishings: chintz for velvet, bare floors and mats for carpets, cool colours for warm. In freak hailstorms, Proust witnessed Mme Swann's traditional spring 'symphony of white with her furniture and her garments', her guelder roses, and her bare arms shivering between 'an immense rectangular muff and cape both of ermine.'

From earliest times, in China and all over Europe, courts and governments with all their attendants migrated from summer country seats to winter residences in town to warm themselves collectively in theatres and ball rooms, in modern Houses of Parliament as in the baths of ancient Rome.

Winter is the natural time for staying late in bed, though means and morals often demand otherwise. Addison, who rose at 2 or 3 a.m. in summer, stayed put until 11 in winter. It took Dr. Johnson until noon to wrench himself from his

11

bliss. There was no household object more comforting to mind and body than the four poster heaped with feather beds, quilts, blankets and coverlets, and closed to the world by canopies and curtains. Our clinically hard and hygienic beds are not abodes.

Company is the oldest and most economical source of warmth. King David, shivering in his dotage, was brought a young virgin to lie in his bosom that he might get heat. But while some people cherished human bodies for warmth, others preferred pets. Liselotte had much more comfort from her dogs than from the Duc d'Orleans, her irritable transvestite husband, brother of Louis XIV. Peasant families were as appreciative of the semi-tropical climate created by their pigs and cattle as we are of our central heating. Many medieval dwellings gained additional heat at no cost from the stabling of animals and storage of fodder beneath a first floor hall or at one end of a longhouse. As the breath of cows warmed the Infant Jesus in the manger, so it warmed Normandy lace makers who worked through the winter nestled in the straw in the cowsheds and poor Auvergne families who moved into the village barn with their beds and belongings. Its legendary sweetness, its constant high temperature and humidity were also considered beneficial to consumptives for whom comfortable retreats were constructed in cow barns.

Solar heating is not so new as we may think. It was first proposed along with solar ventilation, in 1800 by Dr. Anderson, a horticulturalist. His 'contrivance' was a sealed glass shell over the entire south front of the house, up to the chimney top, with valves regulating entry and exit of the trapped air which was heated by sunshine and distributed to the rooms through additional valves. Simultaneous use as a vinery was a bonus. Solar heating first functioned in 1961 in St. George's School, Wallasey, Cheshire. Body heat and electric lighting gave a boost to the temperature just as they do in ordinary homes along with fridges, television sets and other electrical appliances.

3 Braziers and Portable Heaters

Of all heating apparatus, none has had as long and wide-spread use as the brazier. The cradles of civilization – Middle Eastern and Mediterranean – lacking coal and abundant supplies of wood, were warmed primarily by charcoal burned in metal containers.

Charred wood, reduced in bulk and weight, is a more convenient, economical and efficient fuel than raw timber. It is cheaper to transport, easier to store, and longer lasting as much less is required to raise an equivalent heat. Smoke and fumes were eliminated in the open and only the glowing embers brought indoors. Sweet aromas, sought by Greeks, Elizabethans and others, were obtained by adding pinecones or bitumen. The little ash is beneficially disposed of in the garden. Maintenance of the fire is minimal. Narrow-necked Persian vessels called *tennors* needed only an occasional puff through a long pipe. Open basins were stirred with a bodkin, the so-called *marito* or husband that accompanied Italian women of all classes. Fresh embers were carried in, when required, in firepans, causing no disturbance to the company or the heater. Peace, however, could be eternal in an unventilated room warmed by an unattended charcoal fire emitting odourless carbon monoxide. Jovian, Emperor of Rome for only nine months (June 363 to February 364 A.D.) and Philip II, King of Portugal head the list of fatalities caused by braziers left overnight in bedrooms.

Nevertheless, the danger of asphyxiation was no more a deterrent to charcoal burning in the past than it is to gas consumption today. Neither was the embellishment of braziers inhibited by the technical fact that it reduced its thermal efficiency. The plainest iron bucket provides the most intense heat. But the 'brasseries of silver' noted by Evelyn in

13

the Duchess of Portsmouth's apartments, those of bronze, copper or brass with basins and lids pierced, embossed, inscribed, or mounted with ornaments in high relief did more to enhance the sense of well-being [pl.4].

Mobility is the brazier's great and unique virtue. It was useable at any time or place, speedily ignited or conveniently stored away saving the space and money consumed by the construction and upkeep of a stationary fireplace, a stove or a hypocaust. Moderate sized basins were generally set down upon tables or tripod stands [fig.2] and occasionally rested precariously on the lap. Intrepid Italian women often sat with their braziers tucked under their skirts. Larger 'court chimnies', like the 'two rounde pannes made six square wise' for Henry VIII's chamber in the Tower of London, circulated on wheels.

Wealthy Romans relied more upon braziers than hypocausts to warm their villas. Poorer families living one above the other in multi-storey *insulae* had no other source of heat. The Persian *tennor* was placed in a hole in the floor, covered with a wooden lid, draped with a carpet, and used as a table, seat or footwarmer for a cushioned bed made nearby. Throughout Europe, from the Middle Ages to the eighteenth century, braziers alone warmed hearthless closets, and supplemented the heat of log fires in other rooms. Although made redundant by better sealed rooms, coal fires and stoves, they were still used in poor chimneyless dwellings, in better Mediterranean homes to take the chill off rooms when fires were out of season, and in the Houses of Parliament in the nineteenth century. Before the days of the hot water bottle and electric blanket, bed clothes were warmed by a small brazier that was either fitted with a long handle and called a warming pan or was suspended on a bed wagon, an arched wooden frame which when inserted between the sheets suggested a bulbous sleeper and so was known as a *moine* or monk.

The first portable heater to replace the brazier was the paraffin oil stove [fig.3]. This new apparatus, introduced at

Figure 2 (left). Brazier stand. Italian, 15th or 16th century.

Figure 3. Paraffin stove by John Harper of Willenhall, 1890.

the Paris Exhibition of 1878, was an instant success. Count-less halls and passages, churches and assembly rooms were inexpensively filled with gentle warmth and a distinctive odour. Made cleaner and safer, paraffin stoves still retain their economic advantage over all other forms of domestic heating. Until the recent arrival of compact gas cylinder heaters they were the only portable alternative to electricity.

Electric heating commenced in the early years of this cen-tury with a decorative flourish, but with very little warmth owing to the thick glass tubes required to seal the hot ele-ments from air *[fig.4]*. By 1912 non-oxidising elements had been developed, and Mr. C. R. Belling produced the first efficient electric heater, a truly portable fire complete with red glow, toasting rack and trivet for boiling. Modern streamlining disposed of the fireside remnants to secrete radiant bars with convector fans behind discreet vents and dashboards of power controls. With a small apparatus and a large income, instant clean heat can be had wherever, whenever, and for as long as it is wanted.

Figure 4. 'Apollo' electric fire, 1904.

Plate 1. February : Sitting by the fire. Occupations of the Months from the Calendar of the Playfair Book of Hours. French, late 15th century.

Plate 2. January : Feasting. Occupations of the Months from the Calendar of the Playfair Book of Hours. French, late 15th century.

Plate 3. Hand warmer, brass. Italian, c.1600.
Foot warmer, brass, Dutch, 1733.

Plate 4. Brazier, bronze. Spanish, first quarter 16th century.
Brazier cover, copper, Italian, 16th century.

Plate 5. Chimneypiece of painted earthenware tiles from the Palace of Fuad Pasha. Turkish, Istanbul. Tekfur Serai Factory, dated 1143 AH/A.D. 1731.

Plate 6. Pickford Waller. Section of a design for Clarence House, St. James's, showing chimneypieces and radiator (?) under windows. (detail) 1874.

20

4 Open Fires

From the moment it was domesticated by prehistoric man, the fire has been servant and master of every household it has entered – the nourishing, though virgin, mother symbolized by the Roman Vesta; the *enfant terrible*, Loki, of Teutonic mythology. It is a heater, ventilator, lighter and cooker, a mesmerizing entertainer, 'the most tolerable third party', wrote Thoreau. It is also an extravagant consumer of fuel, and a tyrannic demander of attention, tools, labour, time and space. Abuse or neglect it punishes by either filling the room with smoke, setting the house ablaze or going out.

However rigorous the rituals of keeping home fires burning, they were preferable to the difficulty of striking a spark with a flint, igniting dry tinder and raising a flame *[fig.5]*. We, with matches, firelighters, and electric pokers, may not appreciate the virtues of vestal virgins and of curfews, domes of copper or brass, like the one in the Victoria and Albert Museum appropriately decorated with the martyrdom of St. Lawrence, which cherished the embers of one day's fire for kindling the next. They certainly would have been

Figure 5. Prototype model of GEC electric fire lighter, 1950.

appreciated by the maidservant of 1626 who 'stirring be-times, and slipping on her shoes and her petticoat, groapes for the tinder box, where after a conflict between the steel and the stone, she begets a spark, at least the candle lights on his match; then on an artificial fabrick of the Black Bowels of New-Castle Soyle, to which she sets fire with as much con-fidence as the Romans to their Funeral Pyres.' (Nicholas Breton, *Fantasticks*, 1626 cited by W. T. O'Dea, *Making Fire*, HMSO, 1964.) The conflict of stone and steel went on drawing blood from knuckles until 1826 when John Walker of Stockton-on-Tees produced the first friction match. Is it any wonder that spitting on the fire was strictly forbidden, that pissing it out, or allowing it to die was regarded as a curse upon the household?

Hearth and home are significantly expressed by one latin word, *focus*. Wherever the fire is, there invariably is the heart of the house upon which life converges.

Central Hearths

An open fire in the centre of a room is the best social focus, and probably the first. It is certainly the cheapest and easiest fireplace to construct requiring only a shallow pit in the floor, perhaps a low stone platform on which to rest a pot, and a hole in the roof for smoke to escape. Central hearths were particularly expedient for nomadic peoples, and were in com-mon use by Europeans of all ranks and nationalities from pre-historic times to the fourteenth century and beyond. In Westminster School and in Lincoln College, Oxford they continued burning into the nineteenth century, even later in some humble Scandinavian cabins. Although better than re-cessed fireplaces for warming large numbers of people, open fires are less convenient for cooking and more difficult to pro-tect from draughts. They are restricted not only to single storey buildings, but also to wood burning, and are inclined to be dirty. Soot blackened (*ater*) walls gave the Roman *atrium* its name. Until covered with louvred turrets in the thirteenth century, the opening intended to let out smoke also let in the elements. Russian Ostyaks, living underground, were less disturbed by snow than by animals falling through their smoke holes. Tartars were protected by a sheet of ice that formed as soon as the wood was charred, leaving them unventilated and as bleary eyed as the Laplanders punned by Linneaus *Lappi lippi sunt*. Long abandoned for more sophisticated heating apparatus, the central hearth has been revived in America to restore conversation to a hyper-mechanized public silenced by television. Some hang from the ceiling, fire and flue in one brutalist design, others are set like a stage above underfloor ducts, surrounded by an audience in prescribed seats equipped with tables for cook-it-yourself dinners. Paradoxically, central heating has made open fires all the more alluring.

Wall Fireplaces

Like many familiar conveniences the flued fireplace is believed by some scholars to have originated with the Romans. Others disagree, and as there is no tangible proof the debate remains unresolved. Be that as it may, enclosed fires did not come into general use until the early eleventh century when higher standards of living and higher houses with wooden floors militated against central hearths. Instead of being hollowed into the thickness of walls required for defence, early fireplaces were laid on the surface and protected, mantled, by projecting jambs carrying a hood to gather smoke which was not very efficiently drawn by a short flue pierced through the top of the wall and terminating below roof level. As the need for defence receded, so too did the hearth. Although hoods were retained until the sixteenth century and later on the Continent, by the end of the fourteenth century most English fireplaces were flush with the wall. Initially the upright chimney breast was supported by either a flat four-centred arch or a stone lintel or mantel resting on imports or corbels. Enlargement of the mouth to a yawning width of 8 to 10 feet necessitated a longer timber beam, or mantel tree, which was superseded, for fire precaution, by an iron chimney bar. Meanwhile, Near Easterners, ill-supplied with wood, built tall, narrow chimneys to burn logs vertically [*pl.5*].

Cavernous chimneys, their hearth slabs projecting into the room, their walls lined with brick sometimes whitewashed or covered with tiles and fitted with benches, made cosy inglenooks in which members of the household could shelter from the howling draughts caused by the large aperture. There was ample space too for storing the fire irons and cooking utensils, for smoking bacon, drying clothes and herbs. There were loopholes for light and passages to adjacent chapels where prayers could be said in warmth. The large fireplaces of St. Germain gave Louis XIV entry into the apartments of the Queen's maids of honour. Mme. de la Popelinière had a literal fire escape from her Paris chimney into her neighbours. Neither warmth nor adventure was the cause of the inglenook's revival in stove or centrally heated nineteenth- and twentieth-century houses [*fig.6*]. Its vernacular medieval associations were fashionable. Its closed, intimate ambiance, focused on the fire was favoured by Frank Lloyd Wright as a powerful contrast to his open designs looking outward.

With gradual reduction in the thickness of walls, chimney

Figure 6. H. H. Emmerson. Portrait of Lord Armstrong in his Dining Room inglenook at Craigside, Northumberland, built 1870–4 by Norman Shaw. At Craigside a National Trust House.

24

breasts were raised one above the other and projected into the room. This projection, though maintained in England even when flues were aligned side by side in slimmer stacks, was not favoured in France and Italy. From the seventeenth century, as rooms decreased in size and multiplied in number, as the use of coal increased in England, requiring a lower draught and narrower throat to extract smoke, as everywhere standards and costs of comfort rose demanding maximum returns from the fire, the chimney mouth contracted from a yawn to a whisper.

The normal place for a fireplace, at least until the mid-sixteenth century, was an exterior side wall, preferably the side least exposed to wind, below the dais in a medieval hall, and alongside or between window bays. This one-sided arrangement of openings, although advantageous to large painted rooms in Italian *palazzi*, was extremely inconvenient to the internal and external symmetry of smaller classicizing villas, and weakening to the walls, according to Sir William Chambers. Thinking only of the fire, however, Walter Bernan was pleased that in Tudor fireplaces, shaded from the sun, 'the ruddy blaze was always the smiling eye of the room.' Unlike Chambers, he would have liked larger, lower windows allowing landscape and blaze to be enjoyed simultaneously. The opportunity lost in great Tudor palaces, has been gained in the glass ranch houses of America, where the fireplace is so surrounded by view that it might as well be outdoors, and often is for barbecues. Fires provided with circuitous flues could even be fitted *under* windows *[pl.6]*; and, according to George Dance, were not only quite common in eighteenth-century France, but were successfully operating in 1804 in a counting house in the East India Warehouses in London. Although radiators were welcomed under windows, the thought of seeing the face of some passing stranger in the customary place of one's own reflection over the mantelpiece was too alarming, too surreal to countenance.

Situated in a partition wall, the chimney gained a warmer and therefore better drawing flue. It also gave some of its

25

heat to the adjacent room or passage. In the open living areas of modern houses a central pier makes a convenient, compact flue and fireplace visible, in many cases, on two sides. Dutch corner fireplaces, introduced into England by Charles II, were primarily for small closets where unbroken wall space was precious and fireside gatherings either intimate or unexpected. Tiers of shelves laden with china made them more attractive as display cabinets than as heaters.

Mantelpieces

The mantelpiece was, as its name implies, originally a protective cover for a fire made on the surface of a wall. Recessed in the thickness of the wall, the fire needed no further protection and could have been dismantled in the fourteenth century as it was in the twentieth, without impeding its operation. Instead, the mantelpiece was assigned a new and independent role as a decorative cloak concealing the framework of the aperture. The etymological connection between chimney and surround has been traced to the French and Italian custom of hanging wet clothes to dry before the fire. The term, chimneypiece, originally referred to the covering of a chimney breast with a textile hanging, a painting or carved tablet. In modern usage it describes both mantel and overmantel and is synonymous with mantelpiece.

The ornamental treatment of early medieval fireplaces, like other domestic fixtures in that unsettled age, was sparse and ecclesiastical in character *[pls.1, 2]*. Jambs might be treated as coupled columns, stone mouldings, capitals, corbels and spandrels carved with simple geometrical or foliate motifs. Although surviving hoods of the period are gaunt, they may have been made less so by hangings. Queen Eleanor's appreciation of the fire in her chamber at Westminster Palace was doubtlessly enhanced when its hood was painted in 1240 with 'a figuer of Winter as well as by its sad countenance as by other miserable contortions of the body may be observedly likened unto Winter itself.'

From the permanent rooting of previously dispersed life

and wealth in a single residence in the late middle ages chimneypiece embellishment burgeoned for at least five hundred years. Nowhere else, outside of the church, was sculpture better displayed. In marble and wood, steel and concrete, in all sorts of materials, in all countries and centuries the finest artists, from Desiderio in fifteenth-century Florence to Isamu Noguchi in the post-war University of Keio in Tokyo, have performed around fire.

The fifteenth century not only increased the height of the mantelpiece and quantity of ornament, but more important, it introduced a distinctly personal character manifest primarily in heraldic devices and to a lesser extent in portraiture. Assured of attention by the magnetic fire, the chimneypiece became the *mis-en-scene* for ostentatious displays of present wealth and prestige, past heritage and future continuity.

Most of the ornamental and architectural details, some marbles and many of the craftsmen of European chimneypieces from the fifteenth to seventeenth century were imported from Italy. Their overall design, however, was not. Certainly, magnificent chimneypieces, monuments of architecture and sculpture, were to be seen in the Renaissance palaces of northern Italy. Towering chimney stacks of elaborate design were the fame of Venice. But, owing perhaps to the mild Italian climate, the scarcity of wood and preference for mural painting, the fireplace never had the functional or decorative importance there that it had north of the Alps. Improvements in design and construction originated primarily in France and were further developed in England, the home of the hearth.

The most significant sixteenth-century innovation, its precise evolution as yet uncharted, was the two storey or 'continued' chimneypiece rising from floor to ceiling doubling the width of the aperture *[fig.7]*. These towering structures were given pride of place in the best rooms of the best houses where they overwhelmed all other furnishings. Orders, terms and caryatids were freely used to frame and partition panels boasting heraldic achievements, battles won, virtues aspired, pedantic inscriptions, and appropriate biblical and mythological subjects – Shadrach, Moses and the burning bush, Vulcan's forge, etc – all carved in stone, marble, or wood painted and gilded. Later, Flemish influence turned ordered pomp into a nightmarish labyrinth of strapwork and grotesques interspersed with demonic Loki. The extravagance sought by Piranesi to distinguish chimneypieces from doors and windows could have been found in the German mannerist designs of Wendel Dietterlin and in many Tudor and Jacobean apartments.

The architectural character imposed upon the chimneypiece in the seventeenth century improved its proportions, disciplined its decoration, and related it to that of other fixtures, without diminishing its prominence. Columns, capitals, entablatures and assorted pediments – adjusted by Inigo Jones to Palladio's rules, allowed by the French to take baroque liberties – formed aedicules in which the sister arts, painting and sculpture, gave virtuoso performances. Statuary, bas reliefs, cartouches and swags surrounded prized

Figure 7 (left). Chimneypiece of carved wood from Bromley by Bow Old Palace, 1606.
Figure 8. Chimneypiece and hearth of inlaid marble. Fireback, and silver-mounted fire pan, shovel and tongs. The Queen's Closet, Ham House, Surrey. Late 17th century.

pictures. More swags were applied to the walls each side as if to frame a frame. Wealth and distinction, once blazoned in armorial bearings, were now expressed in more general, universal terms, always, of course, for public appreciation.

In private and secondary apartments, even in some great saloons painted or hung with expensive textiles, single storey mantelpieces were preferred. Although 'simple' in stature and profile, rich effects could be created by the use of coloured and inlaid marbles as in the Queen's Closet at Ham *[fig.8]*.

To suit the intimate style of eighteenth-century life, the show chimney was dressed less to awe than to delight the spectator. In the quest for novelty and variety rococo, gothic, Chinese and Moorish concoctions, pseudo-Grecian, Egyptian and Etruscan realities vied with Palladian propriety, the English national idiom *[pl.7]*. Henceforth, until the chimney was exhausted, the numbers of competing styles multiplied and the pace of their succession quickened. Regardless of fashion, politeness, according to Chambers, forbade the representation of nudities and indecencies, of anything capable of exciting Horror, Grief, or Disgust' in the presence of ladies, children and other modest or grave persons. Thus, grotesque Loki was replaced by tamed vestal virgins, except at Dorchester House where even Hercules was strained to support Alfred Stevens's monumental chimneypiece of c.1868 *[fig.9]*.

The high standards attained by eighteenth-century architects schooled in Rome were broadcast to masses of minor craftsmen and their patrons in a deluge of pattern books. By the end of the century the elegant mantelpiece was the *sine qua non* of reception room decor, not only where fires were the prime source of heat, but also in pretentious Russian, German, Scandinavian and American houses warmed by stoves. The unique burnished steel mantelpiece, fender and garniture all mounted with Adam-style ornaments of gilt copper, brass and cut steel forged in the late eighteenth century by the Russian Imperial Armoury at Tula for presenta-

Figure 9. Chimneypiece from Dorchester House. Marble. Designed by Alfred Stevens and installed unfinished in 1868. The figures polished after Stevens's death by James Gamble.

tion to an unassuming English woman, Martha Wilmot, must be one of the most extravagant examples of the folly of international fashion *[fig.10]*. To have been useful, this gift ought to have been accompanied by an asbestos suit and an army of serfs to do the polishing. It was not utility, however, but decorative tradition alone that fixed the mantelpiece on the wall, that keeps it standing over cold hearths, and brings sham surrounds in period styles to hang like picture frames on walls where no fire is expected.

While pattern books, John Crunden's *Chimney Piece Makers Daily Assistant 1766* and others, kept country squires in fashion, enterprising eighteenth-century stone masons, like John Carter, and artificial stone makers, like Mrs. Coade, supplied ready made mantelpieces for immediate installation in the rash of Georgian town houses. Other fabricated materials and simulating techniques gave architect-decorators unlimited freedom to spin webs of harmonious ornament over the chimneypiece, onto the walls, ceilings and every article of furniture including the grate. Economy and expediency substituted stucco for carved wood, painted marble for Bossi inlays, Wedgwood plaques for classical reliefs, Matthew Boulton's gilded cast tin mounts for expensive French ormolu, scagliola for marble, foiled and painted glass for antique porphyry in Adam's overmantel for the Glass Drawing Room of Northumberland House, now in the Victoria and Albert Museum.

Before long the seeds of commercialism planted in the late eighteenth century were yielding bumper crops of machine manufactured mantelpieces of cast iron and wood for large and undiscriminating middle class markets. With increased quantity, quality fell as sharply as price. The down trodden fireplace was seized from the grips of the machine and restored in the 1860s by William Morris, Philip Webb and fellow craftsmen and artists to its original form – heavily hooded and honestly constructed of brick, local stone, and timber. Arts and Crafts vernacular found favour in many smart Victorian homes and went on to have a powerful influence upon modern design. On the other hand, neither man nor machine could long withstand William Burges's hooded medieval giants wielding heraldic arms.

The overmantel mirror was the dazzling successor to the double-decker chimneypiece. Its first appearance in France in the second half of the seventeenth century was prompted perhaps by the fireplace in the mirrored walls of Louis XIV's

Figure 10. Chimneypiece with fender, fire irons and garniture of burnished steel. Made at the Russian Imperial Arms Factory at Tula. Late 18th century.

Petit Apartement at Versailles. To begin with it was incorporated in rather than above the mantel frieze which was raised to considerable height to accommodate another French innovation, the mantel shelf, purposefully designed, according to D'Aviler, to display china in a conspicuous, unreachable position. Shelves were widened and lowered in the eighteenth century allowing mirrors to shoot up as the manufacture of plate glass progressed, in France more rapidly than in England *[pl.8]*. The mirror, as well as a mark of vanity, is an essential emblem of the Age of Enlightenment. It greatly increased the lighting of rooms, to which end it was often fitted with girandoles; it multiplied their size and reduced the real and apparent weight of wall decoration. The resemblance of mirror and Palladian window was a theme much played upon by Adam.

If the chimneypiece need resemble anything, however, then in Piranesi's view it ought 'to be made in resemblance of a cup board or chest of drawers.' A cupboard is precisely what it had been made, not just in resemblance but in reality, by the introduction of seventeenth-century china shelves. As long as there was a shelf, no matter where it was or what it held, the mantelpiece was an article of furniture. In Marot's design of 1710 for a 'Chinese Cabinet', it is literally every inch a display case, visually as extravagant as any Piranesian composition *[pl.7]*, and functionally as busy as a Victorian 'bracket-and-overmantel style' bureau *[pl.9]*. Palladian and neo-classical tastes required the alignment of the shelf to conform to an architectural cornice and its garniture to be symmetrically ordered.

The *garniture de cheminée* began in the late seventeenth century as a suite of vases made in China especially for European mantelshelves. According to the size of one's chimneypiece and purse, one might have three, five or seven pieces, five (three balluster shaped and two beakers) being the norm. Marot's fireplace is covered and even filled with Chinese export garniture. In the eighteenth century Chinese vases were imitated in Delft, Meissen and other porcelains. Addi-

Figure 11. Mantel clock by Vulliamy. English, 1807–8. Given in 1864 to Maria Theresa Villiers by Ernest Augustus II, King of Hanover on her marriage.

tional objects were introduced and the style and contents of the garniture varied to keep pace with fashion. Portable clocks *[fig.11]*, imported at great expense into England and America from Germany, Holland or France and elaborately cased in gilded metal, were precious enough to merit a safe and central position on the mantelpiece. Candelabra were added to light the garniture as well as the company. Suites were specially designed by Adam and his followers, *en suite* with the chimneypiece, the grate, the fender and all the other furnishings of the room *[fig.10]*. Ready made garnitures in the newest antique style could be purchased in basalt from Wedgwood, in ormolu from Matthew Boulton. Sphinxes, although highly fashionable, were less endearing than Staffordshire dogs who accompanied their fire-loving British masters as far afield as Patagonia. Early in the nineteenth century B. Day of Birmingham was manufacturing patent chimneypiece ornaments in the form of telescoped gothic pinnacles 'so constructed that they may be used for fire screens, flowers or scent jars, time piece containers, candle sticks and various other purposes'. Dressed by ingenious Victorian ladies the mantelshelf wore a valence of velvet, hand painted, embroidered and fringed, under an 'artistic' array of bric-a-brac, precious plates, cheerful flowers, sentimental letters, photographs and cards crowding the face of the inevitable clock.

Forgetting Marot's army of shelves and brackets, Victorian decorators like R. W. Edis puffed themselves at having transformed a 'mere ornament' of a chimneypiece into an 'absolutely useful piece of furniture.' Marble surrounds were sent to the attic to make way for wooden china cabinet, gun cabinet, or dressing table mantelpieces which could if necessary be moved to a new house. But the ultimate in chimneypiece furniture is Gio Ponti's 1950s walnut corner cabinet with doors opening to reveal a fire and garniture.

Looking down from pseudo-medieval and Renaissance heights, some Victorian architects saw no use whatsoever in a mantelshelf that had sunk so low as to appear as a seat. Would they, one wonders, have stepped or sat on Timo Saponova's 'James', a streamlined, stainless steel octopus fireplace with a barbecue hot plate over the grate, built in tables, seating, lighting, air conditioning, stereo system and TV? [fig.12].

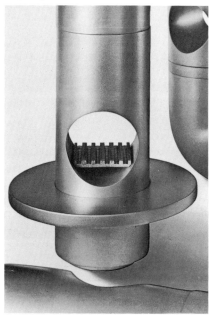

Figure 12. 'James', the complete servant. A stainless steel tubular fireplace adjustable to any shape or size, usable indoors or out. Designed by Timo Sarpeneva of Finland, 1973.

Figure 13. Andirons incorporating candle holders. French, 16th century.

Tools and Paraphernalia

Andirons (fire-dogs, *chien de feu*), in pairs supporting the ends of logs to prevent rolling and to allow air to pass underneath, are the oldest of domestic fire implements, the earliest examples in this country dating from the Iron Age. In central hearths and in medieval fireplaces stood massive iron fire-dogs ringed and bracketed for spits and utensils, topped with cup-shaped cressets for lights or warming pans [*fig.13*]. Between them came low creepers (dog-irons, or *chenets*) to lessen their burden. From the sixteenth century, fire-dogs, like hearths, grew smaller, more numerous and ornamental, bearing emblems, beasts, and arms. At Hampton Court, Cardinal Wolsey had no less than forty-seven pairs of andirons, one more elaborate than the other. Bronze gods and goddesses by the foremost Renaissance sculptors enobled Venetian fires [*fig.14*]. There were seventeenth-century andirons of intricate repoussé brass and coloured enamels

Figure 14 (left). Andiron. Adonis with satyrs and cupids. Arms of Daniel Barbaro, Venice. One of a pair, the other of Venus. Bronze, Venetian, c.1570.

Figure 15. Andiron of cast brass with enamel decoration. One of a pair with Stuart arms. English, c.1680–9.

[fig.15]. There was silver for Restoration nobility, and of course for Louis XIV; but not for their eighteenth-century successors who made do with gilt bronze or cast iron dogs of rococo, gothic, neo-classical or Chinese design. When rendered redundant by coal, andirons became the ornamental supporters of dog-grates. Fire-cats supported plates and pans to warm before the fire.

Irons required to tend a wood fire were a strong pair of tongs for grasping logs and embers (smaller brand-tongs lifted live charcoal for lighting candles or rush lights); a long handled poker, billet hook or two pronged fire-fork, not to poke but to shift logs forward; and a shovel to remove ash, pierced to sift coal cinders. On the whole these implements were simply designed for hard wear. The set of silver mounted irons, made in 1673 for the Duke and Duchess of Lauderdale on the occasion of Catherine of Braganza's visit to Ham House, are exceptional survivals of what was presumably to be seen in royal state apartments *[fig.8]*. Fashionable eighteenth- and nineteenth-century fireplaces might display one set of polished brass or steel tools for beauty and another for use, which to an American visitor like Stephen Fiske was an inevitable source of confusion. Some irons reclined on low rests, others stood upright in the arms of ornamental stands equipped perhaps with a kettle stand and occasionally joined to the fender.

The poker was the coal age weapon *par excellence*, required by Bernard Shaw's *Man* (but not Superman) to master his wife as well as his fire. This emblem of male supremacy was traditionally reserved for its owner, and shared only with long-standing friends. Let Stephen Fiske's 'foreign hand touch the poker and the fire grows sullen and dies out.' The disc-poker, invented by T. Pridgin Teale in 1888 on the principle of the sugar crushers used by toddy drinkers, was intended for flattening the surface of coal to check its fierce burning and extravagant consumption encouraged by his improved grate. Economy continued to decrease the rate of combustion and with it the size of the poker. By 1912, this once formidable implement was a 'mere toy.'

Firebacks *[fig.16]*, produced in Kent and Sussex Weald

Figure 16. Fireback. Hollandia and the lion of the United Provinces. Dutch, 17th century.

37

foundries from the end of the fifteenth century, were used to preserve the brickwork of the hearth. Crude patterns, made by pressing available objects – ropes, shells, swords, etc – into the open sand moulds before casting, were gradually replaced by more sophisticated compositions of heraldic, natural and narrative subjects specially carved in wood for the purpose. As the width of the hearth contracted, the slabs became narrower and taller. Incorporated in coal grates, their ornamentation was simplified and generally restricted to side panels where it was less likely to be caked with soot. Although some of the heat absorbed by iron firebacks was reflected into the room, much of it was lost up the flue. For that reason they were abandoned and replaced by slower absorbing firestone bricks or tiles.

Fenders or curbs of stone enclosed a few medieval hearths. But the familiar metal guards used to protect the floor and hearth rug from burning coals were not common until the seventeenth century. Most wood fires required no more than andirons and a raised hearth stone to contain them. Nor did attractive marble slabs, like the one at Ham *[fig.8]* need masking. It was mountains of coal heaped in high grates that needed fending. From architects and cabinet-makers, ironmongers and chimney furnishers, fenders could be had in every fashionable style, in burnished steel or brass, silver for the great and cast iron for the multiple *[fig.10]*. Some, especially in the early eighteenth century, stopped at or were hooked to the jambs. Others spread around the entire chimneypiece and beyond. There was even a circular fender that moved back and forth in a circular groove round a circular hearth. Although most fenders remained low, many in late Victorian times rose to seat height and were upholstered for that purpose. Club fenders provided seating on three sides. Seat curbs were less sociably padded only on the ends where cupboards were often fitted, one side for coal the other for slippers. Not only were fenders required to contain the burning fire, they were made by Alexander Walker in 1777 to encourage it by blowing fresh air at it through holes pierced in the hollow rail originating outside the house; by Mr. Blundel to sift cinders from ash in a splayed grid. For summer, Mrs. Panton recommended a fender of 'virgin cork . . . filled with moss and then jam pots stuck in it full of flowers.' Housewives who treasured their skirts and despised dirt followed the advice of R. W. Edis and got rid of their 'artistic' fenders with treacherous curves for plain polished marble or stone curbs.

. . Fireguards of wire mesh are not a modern safety precaution for children *[fig.17]*. In 1767 Mrs. Lybbe Powys saw 'lacquered wire fireboards' in front of every chimney in Buckingham Palace, so fine that 'even the smallest spark cannot fly through them while you have the heat and they are really ornamental.' You would have much more protection but less heat from William Clark's glass-window-guard which, except for a moveable pane to admit the poker, eliminated all communication between room and fire, air being introduced

Figure 17. Chimneypiece designed by Thomas Hope for his brother, Adrian, at number 4 Carlton Gardens, Pall Mall. Marble and iron with ormolu mounts. Brass basket grate with wire mesh fire-guard. 1837-46.

by tubes from outside. Satanic stoves were sometimes imprisoned behind tall cages.

Coal scuttles [*fig.6*], boxes or hods are recorded in the eighteenth century, Lady Grizell Baillie had one of copper in 1715, but no examples prior to the nineteenth century survive. Many were extremely attractive, made of polished brass, or painted and lacquered in all styles, with their contents secreted under helmets and lids in removeable sheet metal containers in Mr. Purdon's purdonium. Nevertheless, they were not considered presentable in the very best reception rooms. There coal was brought up from the cellars in special coal lifts, wheeled to a back passage in coal trolleys [*fig.18*], and carried into the fire when summoned by the fireside bell. The enormous logs burned in the past, unlike our foot sized billets, required no storage containers. Edith Wharton's cultured taste ran towards gilded and painted Italian cassoni for use as wood boxes. Rush and wicker baskets are much more practical and familiar even at noble firesides [*fig.6*].

Bellows, since antiquity, were required to encourage sparks to flame. The modern form of nozzled leather bag within

Figure 18. Coal wagon at Ham House, Surrey. Late 19th century.

40

wooden frames was in use in the Middle Ages. From the fifteenth century onwards, bellows were elaborately enriched,
the frames carved, painted, lacquered and inlaid, mounted
with bronze, or at Ham with embossed silver incorporating
the Lauderdale monogram. Such precious possessions had
ornamented cases in which they hung on the chimney jamb.
The French, instead of bellows, preferred more elegant fire
fans which could be used to enliven a dull fire or to cool an
overheated face *[pl.10]*.

Ventilating fenders we have already noted. Eighteenth-
century hand operated fans, re-introduced and patented in
the nineteenth century, were more ingenious than labour
saving, and were inclined to spread the ash more than the
fire.

Fire cloths, ordinarily of leather, of damask or tapestry in
better houses, were commonly suspended from medieval
mantels where they performed the smoke screening task of
banished hoods. By temporarily contracting the mouth of the
fireplace, they increased the draught up the flue, thereby
helping to start the fire and remove its smoke. Most fire cloths
were either hooked in place when needed, or permanently
hung, like curtains, on rods. Some were contrived as Venetian
blinds to be lowered to an expedient position and dropped to
the floor when the fire was not in use. Prince Rupert's improved fireplace of 1678 replaced the fire cloth with an iron
door hinged to move back and forth. Such draught regulating
devices, some operating like sash windows, were incorporated
in many eighteenth- and nineteenth-century 'scientific' fireplaces. Curtains, instead or in addition, were favoured for
'artistic' Victorian effects *[pl.9]*. Modern wire mesh blinds
and curtains have been used since the early nineteenth century for the dual purpose of increasing the blaze and safe-
guarding the room.

Fire screens prevented fireside sitters from being blinded
and scorched in front, while tall screens and high settle
or chair backs kept the rear from freezing. Mobility is the
essential theme of the fire screen upon which there were

innumerable variations in shape and embellishment. Hand screens of wicker or wood were commonplace, even in noble households, since the Middle Ages. One of richer design is depicted in use on the Brussels *Winter* tapestry of c.1700 *[pl.10]*. Standing screens were ordered in *The Booke of Curtesye* of 1440 to be provided by the Groom of the Chamber to protect his lord from the heat when he was seated at table. Panels mounted on tripod poles, or iron 'screen sticks' as described in the Ham House inventory of 1679, sufficed for wood fires and remained fashionable despite their inefficiency against the intense heat of coal grates. Larger, four legged cheval screens with fixed or sliding panels were sensibly preferred in the late seventeenth century when coal was admitted into the best apartments. More generous protection was subsequently offered by unfolding cheval screens, horizontal sliding panels and swinging leaves. However mounted, whatever their size and shape, firescreens were always a fertile field for decorative painting, inlay and needlework, even for the taxidermist. As aggressive pokers were for men, so protective screens were for women whose lives were largely spent in the sedentary occupations of fireside chatting, sewing and reading, whose layers of clothing would have made their comfortable position hellish. Byron, had he seen through Lady Holland's dress the 'wadded gown' which was the only kind of heat she liked, might have understood why she always had 'that damned screen between the whole room and the fire.' She was warm, but he who '. . . had never yet found a sun quite done to his taste, was absolutely petrified, and could not even shiver. All the rest, too, looked as if they were just unpacked, like salmon from an ice-basket, and set down to table for that day only.'

The furniture that custom rather than design brought to the fireside for comfort and convenience has altered little since fireplaces began. Draught proofing settles *[pl.2]* fitted with foot rails and convertible into tables, became well padded settees, high backed and winged easy chairs *[fig.6]*.

The rails became independent foot or so-called fender stools *[pl.10]*. Hearth rugs appeared as early as the fifteenth century to keep feet from freezing and floors from burning. Separate tables were drawn up for games, informal meals and later for tea. Their place is now occupied by coffee tables where prestige and fashion are conspicuously displayed according to fireplace tradition. This intimate, sitting room arrangement was often maintained summer and winter in front of real and electric fires. It did not pertain, however, to public reception rooms.

One of the few articles of furniture specially designed for the front of the fireplace was the horseshoe shaped wine table or 'Gentlemans social table'. This late eighteenth-century innovation was for after dinner use in an age when liquor was not consumed during meals. Topers, serving themselves from bottles ranged in a well or in moveable coasters attached to a metal rod, must have welcomed cooler tables fitted in back with folding firescreens.

Chimney boards closed the hearth in summer to protect the room against rain and soot falls. They appear in fifteenth-century Flemish paintings as plain wooden panels making the tall mantel indistinguishable from the door. Great apartments in Italian and French palaces had chimney doors of metal, richly embellished and hinged to allow use of the hearth as a garde robe, a refuse bin and a convenient hiding place for indiscreet visitors. Henry II's queen and mistress were the more chaste for locked chimney doors. Inexpensive wooden boards invited Marot's Baroque and Adam's Etruscan designs harmonizing with the decor of their rooms, Mrs. Delaney's collages of coloured papers and paints, and paintings by amateur artists. Fashion, however, was the bane of nineteenth-century hygiene. To board a chimney was, in the opinion of Lewis W. Leeds, 'as absurd as to take a piece of elegantly tinted court-plaster and stop up the nose, trusting to the accidental opening and shutting of the mouth for fresh air – because you thought it spoiled the looks of your face to have two such great ugly holes in it.' Healthy hearths held unfolded Japanese umbrellas or carefully folded paper fans raised in the grate on mountains of pinecones.

Coal Grates and Smoke

While mantelpieces performed a facile show on the wall, hearths laboured against mounting odds to satisfy incessant demands for efficient, economic and clean heat. Our present energy crisis, if it be any consolation, has good Renaissance precedent. The rise in living standards throughout the fifteenth century, demanding on one hand more houses, ships, tools and furniture made of wood, on the other arable produce from depleted forest land, produced, by the mid-sixteenth century, a shortage of fuel for the fireplaces it had multiplied. Despite rocketing costs and cries for conservation, the answer then, as now, was not less fuel or worse (peat, dung, or straw), but new fuel. In 1577, William Harrison, watching 'woods go so fast to decay, saw the use of coal of which there was 'such plenty . . . as may suffice for all the realm of England . . . beginneth to grow from forge into kitchen and hall.' Although in towns its growth was rapid, in the country in general and in rich houses in particular another century was required to break down social prejudices that forbade its use outside the entrance hall, in the company of ladies, tapestries and food. On the Continent, where supplies were not so plentiful, its use was longer delayed and even more restricted. Seacoal, so-named for its mode of transport from Tyne to London, certainly raised England's indoor temperature. It also raised problems of unforeseen magnitude.

The first requirement is an open container of iron raised to admit sufficient air to keep the coal alight. Probably from braziers placed on hearths the simple basket grate on feet and dog grate incorporating andirons and fireback developed. Both of these were moveable. The first to be built-in was the hob grate with a basket flanked by side casings replacing earlier masonry supports [*fig.19*]. Hobs, in addition to improving the efficiency of the fire by raising the grate and

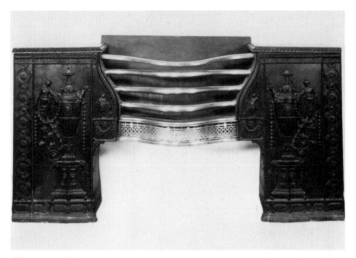

Figure 19. Hob-grate from Hamilton Palace. Possibly designed by Adam. Made by the Carron Company, c.1823.

reducing the opening, were extremely useful for warming food and drink, and, when fitted with front doors, served as ovens. The term, hob nob, is said to derive from the customary offering of beer warmed on the hob or cold from the nob or table. Screened by a metal arch to discourage smoke, the hob grate became a Bath grate of enormous popularity, despite its extra powerful and wasteful draught. Basket, dog, and later hob were treated like their ancestral andirons, to a variety of ornamental embellishments. Around the tough grid were finials, standards, aprons, firebacks and hob fronts all dressed according to fashion in Chippendale's Chinese, Ince and Mayhew's gothic, Adam's neo-classical, Alfred Stevens's neo-Renaissance, or Charles Mackintosh's tendril *art nouveau* styles to match their mantels [*fig.20*]. The high quality cast iron grates manufactured in the last quarter of the eighteenth century by the Scottish Carron Company [*fig.19*], in which John Adam was a partner, inspired wholesale imitations all over the north of England. By the mid-

Figure 20. Fireplace from the Willow Tea Room, Glasgow. Iron grate, blue and white tile surround. Designed by Charles Rennie Mackintosh, 1904.

nineteenth century, mass produced iron fittings – grate, hobs, back and sides all in one – could be purchased cheaply from local ironmongers and easily installed in the waiting gap. The poor were content with economy and availability, the middle classes with 'artistic' appearances. Few were concerned with efficiency. Nevertheless, the cast iron grates, in colour at least, were much more suited to their dirty task than the 'silver cradell' presented in 1613 to Carr, Earl of Somerset, or the dazzling 'jewel grates' of burnished steel mounted with ormolu favoured by Robert Adam.

Coal fires, when introduced into capacious wood burning chimneys, were difficult to ignite and free to send their smoke and fumes into the room. Yellow coal tinge on tapestries and faces was as feared in the sixteenth century as black plague. Paradoxically, the problem was continually aggravated by better built houses with better fitting doors and windows, planning improvements, and high aesthetic standards of architectural design. Reducing the size of rooms, reduced the draught and in so doing caused fires, even of wood, to smoke unless doors or windows were open. Thus, airing the fire meant freezing the company. Tall chimney stacks, that distinguished great Venetian, Tudor and Vanbrughian houses, embarrassed the tidy profiles of neo-Palladian villas. Lacking classical provenance, they were cut short and concealed, weakening efficiency to strengthen beauty. Well insulated, over furnished Victorian houses would have been uninhabitable had the folly survived. Chimney stacks returned in multitudes to darken England's skyline.

In the way of all nuisances, smoke was suffered by the silent majority and fought by a scientific minority. Smoke doctoring, by the eighteenth century, had grown to a thriving profession. Its Hippocrates was Sir Hugh Platt who was the first, in 1584, to reduce the mouth of the fireplace and to concoct smokeless 'cole balles', an invention that may have existed earlier in China, and was repeated by Evelyn and others. The first of many convection fires (introducing air to convey the heat into the room) was perfected in 1624 by a French medical doctor, Louis Savot, from an apparatus he had seen in the library of the Louvre. Savot's fireplace had a tapered throat and contracted mouth faced with radiating iron plates forming ducts through which air was drawn in from the room and returned by a vent in the mantel. One of these iron clad fireplaces placed back-to-back with another whose flue was closed could heat two rooms. Prince Rupert's fireplace of 1678 promised to extract the smoke without losing its heat. After igniting the fuel, which he seems to have made more difficult, an iron baffle was drawn forward to the front half of the flue, forcing the smoke down before it rose up the rear half.

Feeding coal fires with air at the correct rate and place depended on there being sufficient air coming into the room and the occupants not being disturbed by its draught, neither of which could be guaranteed. One remedy was to introduce air from outside. Attempts were made by Sir John Winter in

1658 and by Glauber adapting the Philosophical Furnace used to make his famous liver salts. The decisive caliduct fireplace was conceived by Cardinal Polignac and made public, under the pseudonym Gauger, in the first treatise on fires, *Le Méchanique de Feu* of 1714. Polignac splayed the jambs of the mouth and faced them with metal reflectors behind which air, brought from outside, was heated in several chambers before being expelled through caliducts. The hot air could be distributed not only to adjacent rooms, but, by flexible tubes, to upper chambers and between the bed sheets. Polignac's plan was championed by Dr. J. T. Desaguliers who published an English translation in 1715, with adjustments for coal burning. The English distrust for burnt air smelling of iron, which caused them to oppose stoves, was further hardened by the death of a bird in the course of Desagulier's demonstration of the safety of his apparatus. The foreign invention was rejected. Its many imitations, Benjamin Franklin's among them, had little more success.

Coal fires, however aired, were inclined to be sullen and smoky when topped-up with fuel. To deal with such misbehaviour, there was a host of smokeless grates offering fuel from below. Franklin's rotary grate was replenished on top and then turned upside down to burn, whereupon it could be pivoted to face each member of the audience in turn. To this, nineteenth-century inventors added a variety of mechanical stokers which hoisted coal up to the grate or lowered the grate to the coal. Most of these patent smoke cures, many of them demonstrated at the Smoke Abatement Exhibition of 1882, were more ingenious then efficient, and far too expensive, unattractive and incomprehensible for the ordinary fire lover.

The 'apostle of fire side comfort' was Count Rumford, Benjamin Thompson (1753-1814) of Rumford, New Hampshire, an Anglo-American soldier, politician, and scientist, a reformer *par excellence* of armies, beggars, catering, cooking appliances and fireplaces. Operating on the dual principle that heat is a form of energy, which he was apparently the first to recognize, and that virtue is a by-product of happiness (rather than the reverse), he invented new kitchen ranges, precursors of the familiar Aga, that conserved energy, improved the taste and reduced the cost of cooking. This brought him to the fireplace where he directed flues and throats to be lowered and narrowed in order to increase the draught, and all iron to be banished except for the grate which was brought forward between splayed jambs made of firestone tiles or bricks that did not rob so much of the heat. Throughout England, starting from Lord Palmerston's drawing room and spreading to countless modest parlours, fireplaces were 'Rumfordized'. For appearance, if not always for efficiency, splaying and tiling became the height of fireplace fashion. To the increasing demands for economy, especially from the lower classes, the answer was Dr. Teale's 'Economizer' of 1884 which simply added to Rumford's design an ash pit under the grate, with an adjustable shutter to reduce the draught. Coal, instead of being rapidly consumed, could burn here for up to twelve hours unattended.

Unattended for far longer, for near four hundred years, the refuse of our comfort, the smoke and soot caused by incomplete combustion of coal was discharged onto our streets. Complaints were legion from the sixteenth century onwards, but to no avail. John Evelyn's *Fumifugium* of 1661 against pollution was abortive, so too were bizarre nineteenth-century proposals to draw city smoke into sewers and issue it from towering stacks to the countryside. Scientific measurements of the effects of our indoor warmth on our outdoor climate began in 1887 to tell us in numbers what we had long seen, what Henry Meister, a Frenchman in London in 1789, had expressed better in words, that Vulcan had defeated Apollo. We were sunless but cosy. 'The vapours of sea coal acting on nerves and membranes' had, as Meister predicted, destroyed our sensibility and made us phlegmatic. So numbed by warmth were we as to consent to employment of children to sweep our chimneys. It took one hundred and two years for social reformers and novelists like Dickens and Charles

Kingsley to bring this scandal to an end in 1875. Likewise, London smog, though aired in Parliament, fought by the Smoke Abatement League, bottled and sold to tourists, had to claim the lives of an extra four thousand elderly people, and prize cattle at Olympia in December 1952 before the burning of coal was banned by the Clean Air Act of 1956.

To keep home fires burning, the National Coal Board developed a 'smoke eater', an extra combustion chamber which laundered the smoke delivered to it by electric fans. Although this protracted the use of cheaper grades of coal, the cost and trouble of installation could probably buy a more efficient heater. Coke presented ignition problems solved easily in the 1950s by packets of firelighters or electric pokers, with more difficulty in 1882 by Dr. Siemens' Gas and Coke Fire. By the 1870s solid fuel could be dispensed with, or imitated in metal lumps, and gas burned instead in 'artistic' grates lined with asbestos [fig.21]. Neither smell nor risk was so prohibitive as the cost which was equated in 1874 to 'open fires fed with spirits of wine.' The same has been said of Gas Log Fires presently in vogue, with the further complaint that the realism of their indestructible logs and mock ash invite unwelcome feeding of waste paper and cigarette stubs. Pokers are not required, but flues are. Only electricity can bring, as it has done since the 1920s, the magic glow of a fire in its traditional surroundings, without a chimney and without heat if so desired.

Labour and Love

The drudgery forced upon Cinderella by coal fires raging throughout the home makes a sinecure of the medieval hearthman's labour of lugging long-burning logs to a few hearths. Daily, for half the year, she plodded invisibly from room to room to lift the hearth rugs, sweep and sift the ashes in her portable cinder pail or *tantiseur-étouffoir*, rub and blacken the grates, polish the paraded paraphernalia from fender to poker, scrub the hearth tiles, the slab, and the mat on which the coal scuttle sat. The mantelpiece, if it was stone,

Figure 21. Gas fire with tufted asbestos radiants, 1882.

she 'daily cleansed', according to John Wood of Bath, 'with a
particular white wash which . . . rendered the brown floor
like the starry firmament', adding to her work. Wood and
marble mantels were a blessing for which she was taxed by
the garniture groaning above. Having perfected the setting,
she fetched the coal, kindling, and waste paper, systematically
laid it over the cinders on the grate, and fought with the flint

51

or the sulphur match to light it. The task, far from ending, had just begun.

In exchange for its extra heat, the coal fire demanded full time nursing. Its frequent sulks called for skilful poking which brought forth a spray of ash for Cinderella to remove with her brush and pan when she replenished the coal scuttle. This she might have done up to ten times in one room and the same in others, for, according to the calculations of Frederick Edwards, Jr., author of *Our Domestic Fire-places*, 1865, an ordinary grate was stirred twice as often as refuelled and thus attended to a total of at least twenty or thirty times daily. Before going to bed, she raked the dying coals, put the fire-guard in place, and laid her wood on the hob to dry, ready for the whole procedure to begin again at dawn. Cinderella, though occasionally too generous with coal, could always be relied upon to satisfy the needs of the hearth. As long as there were thousands like her in service, the sitting room fire remained an English institution.

Like all institutions, however unsatisfactory or outmoded, it was defended to the last. Several rational arguments were voiced in its favour. It was the only form of domestic heating that everyone, ordinary servants especially, could understand; that provided wholesome ventilation for the room, cool air for the lungs and warm for the body, albeit half at a time; that allowed each individual the freedom to shiver or sizzle as he chose. Moreover, no steam pipe or hot air duct had the power of its colourful, crackling flames to distract the mind and entertain the senses. Sentiment, prejudice, associations of all sorts play an important part in the survival and revival of the open fire. The symbolism of home, of hospitality, heart and soul is inviolate. To those who kept it burning, it was a sign of social status, *savoir-vivre*, and established wealth better than new American riches. To others equipped with central heating, it was a mark of poverty or, at best, conservatism. It was one of the 'remnants of despotism still lying dormant in the male breast' that hindered helpless housewives of 1913 from the emancipation of gilded radiators.

Plate 7. Giovanni Battista Piranese, Egyptian style chimneypiece. Engraving from *Diverse Maniere d'Adornare i Cammini*.
Rome, 1759.

Plate 8 (left). Chimneypiece from the Music Room of Norfolk House, St. James's Square. The marble mantel attributed to Giovanni Battista Borra; the carving of the overmantel to John Cuenot, 1753–6.

Plate 9. Pickford Waller. Design for a curtained chimneypiece with bracket and shelf overmantel. 1870s.

Plate 10. Winter. Tapestry after A. van Schoor. Brussels, c.1700.

5 Stoves

Open fires pale before closed stoves dispensing four times as much heat but no smoke, demanding less fuel, no Cinderellas or sweeps. By shutting the door on a fire in a container, giving it a limited supply of air and a controlled exit for smoke, its rate of combustion, its consumption of fuel, mainly wood, the cost and labour of replenishment, the loss of unused heat up the flue are all reduced, as is the draught through the room. The longer combustion gases are retained, the more their heat is absorbed by the brick or earthenware casing. This is slowly, uniformly conveyed to the air over an extended period of time, as it is by a night storage heater with electric cables embedded in concrete. Obviously, the larger the volume and surface of the stove, the more heat it returns. In Russia, Germany and Scandinavia magnitude was essential.

The giant stoves that mastered icy climates since the Middle Ages, and still do in some regions, were probably derived from primitive ovens, and are upright relations of Roman hypocausts and Chinese *kangs*. Although it was customary in northern Europe to live in the company of the enclosed fire, it is not essential to do so in order to benefit from the intense heat of its gases. In China stove living was a mark of base poverty. The very rich banished their furnaces outside the house, the middling to an adjacent room. Romans concealed theirs under the floor. The smoke was made to circulate just as it is in a standing stove, but through longer channels – under the entire room in a hypocaust and *ti-kang*; under a platform, a *kao-kang*, which served as sleeping and living quarters; or vertically in the thickness of the walls in some hypocausts and in the *tong-kang* which Chambers adapted for the conservatory at Kew. To prevent gases from escaping and to moderate their heat, the surface of *kangs*, hypocausts and stoves was overlaid with thick earthenware tiles reinforced by transparent or coloured glaze. Not only did such insulation ensure physical comfort, it provided aesthetic pleasure too.

In Germany, Austria and Switzerland stove construction resulted in the development of a special class of ceramic manufacture – hafner ware – which flourished from the fourteenth century onwards. Pottery tiles modelled in high relief and coated with monochromatic green lead glaze – like those depicting the Triumph of Mordicai on the stove made in 1578 for the Convent of St. Wolfgang in Baden by Hans Kraut of Villingen in the Black Forest – were the principal product of German hafner, or ceramic stove makers. On Swiss stoves manufactured at Winterthur from the sixteenth to eighteenth century by the Pfau family and other potters these tiles were often combined with flat faience panels freely painted in rich colours on tin glaze. In addition to stoves, the hafner were responsible for making architectural accessories (wall fountains and tomb slabs), water, wine and cider vessels, dishes and ornamental plaques that were esteemed among the finest products of the German ceramic industry.

As house building and furnishing improved from the seventeenth century onwards, stoves, though still large, were better proportioned to their surroundings. Refinements in the manufacture of pottery and porcelain permitted a wider variety of shapes and ornaments. Lady Mary Wortley Montagu marvelled in 1716 at painted and gilded Austrian stoves 'in the shapes of China Jars, Statues, or Fine Cabinets so naturally represented they are not to be distinguished.' To Mark Twain a century later the 'stately porcelain things that look like a monument' brought chill thoughts of death.

Temperatures and embellishments alike rocketed with iron stoves introduced in France in the late seventeenth century, perfected in the eighteenth century by Benjamin Franklin and others, and finally mass produced. Emitting heat thirty times faster, they could be considerably smaller than their brick brothers, and were eventually sent to the basement to heat the whole house. Casting answered, economically, all the capricious tastes of an expanding market. There were stoves designed by Robert Adam and James Wyatt cast as Roman urns [*fig.22*] or as pedestals supporting candelabra or lamp-bearing figures. The medieval knight in armour designed by Sir John Soane in 1799 to warm the Marquis of Abercorn at Bentley Priory is a predecessor of C. J. Richardson's 1870 'Baronial' ventilating stove with a helmet that opened to receive water and pikes to control the draught. There were also palm trees and Indian temples suggestive, if not productive, of tropical warmth.

A stove heated room in northern Europe had the same name and requirements as its heater; as little incoming air as possible, and an exit for what is vitiated. Draught proofing was mainly effected by sealing windows and closing doors. But this, in a snow bound cabin, was not sufficient. To gain additional heat, Russian peasants, three generations together the whole winter through, performed all the functions of life

Figure 22. Cast iron urn stove from Compton Place, Eastbourne, Sussex. Possibly designed by Adam. Made by the Carron Company. Late 18th century.

on the flat top of the stove, or on a wooden scaffold erected over it, using the lower part for cooking. More genteel persons, like the nuns at the Convent of St. Wolfgang, had seats built into or around their stoves. Ventilation, unlike insulation, was thoroughly neglected. Foreign travellers entering Russian and German stoves were invariably stifled by the heat and fug, and nauseated by the stench of 'children set upon their stools', wet clothing and boots drying, 'rammish clownes' sweating, victuals stewing, all on the stove. On departing, they were followed by an issue of steam 'like smoke from the crater of a Volcano.' Stoves were held by them to account for the boorishness of the Germans and for all sorts of Russian characteristics: foul humours, scurvy, addiction to drink, aversion to physical exercise, preference for chess, cards and other indolent amusements, the hot-house beauty of some faces, the dark hue of others tanned by the heat and mended with red and white paint. Purge was essential to stove dwellers, hence their frequent submission to even hotter steam baths or saunas, followed by an exhilarating plunge naked in the snow.

Life with a red hot iron stove could be just as insalubrious. To many Englishmen it was hell. The air they found too hot and too drying. Particles of scorched dust produced an offensive and irritating smell. Stove malaria and iron cough were new Victorian diseases. There were further complaints of constipation, chest pains, throbbing of the temples, vertigo, fullness in the head, confusion of ideas and cold feet. Stove doctors offered innumerable remedies to subdue the demon. Dampers were commonly applied, to be operated manually until 1849 when an American, Elisha Foote, patented the first automatic thermostat. Dr. Guerney, in 1882, administered fins to the surface to dissipate its heat. The Gill stove did the same. Others, following Cardinal Polignac's cure for smoking chimneys, brought in fresh air to be warmed in an outer cylinder before being dispersed through artistic perforations [fig.23]. In vain, many attempted to quench the thirst of the stove with pans of water. To bury the stove, as the Romans did the hypocaust, might have solved the problem, but it infringed upon the democracy of having a heater 'under the management of those whose feelings were to be consulted.' An acceptable compromise was to isolate it in a hall. So heated were the complaints of the clerks at the Bank

Figure 23. 'Italian' air stove. Bronze and brass with panels of blue and white earthenware by Minton. Designed by Alfred Stevens and exhibited at the Great Exhibition, 1851.

of England that in 1787 all stoves were removed from the building.

Most of the faults blamed on the heater were due to the people it heated who abused the stove by overfeeding it, spitting on it to test its heat, and by failing to understand its operation, yet placing it in the care of even less informed servants. The hot-headed were quicker to dispense with or doctor the heater than to dispense with the vitiated air of their rooms.

To gain the heat of the stove meant relinquishing the light of the open fire. In the days of candles and oil lamps, this was a considerable sacrifice. Although willingly made by those struggling to survive in sub-zero temperatures, it was not so acceptable to others living in better climates or to the few whose pleasures were intellectual. Sebastian Mercier complained in 1788 that 'the sight of a stove extinguished imagination.' Frederick the Great's preference for 'the cheerful and vivifying effect of a fire' was taken as proof of 'how superior he is . . . to the generality of German Princes, who never *see* a fire and are satisfied with *feeling* its operation.' To most Englishmen seeing was believing; the idea and appearance of warmth was as good as the substance and more economical. Objections to the gloom of the stove must have been sufficiently widespread for it did not take long for a compromise to be found. The eighteenth-century Franklin stove and its late nineteenth-century progeny of 'open stoves' had a proportionally larger, centrally placed door which could be left open for viewing and closed for the proper business of heating. Fires behind heat-resistant glass, available from the 1870s, were less illuminating but warmer. As a further concession to fire lovers, most open stoves, Franklin's in particular, were designed to be placed in a traditional hearth, thereby preserving the look of the mantelpiece and its use as a display case.

The 'red hot demon', despised by Dickens and gingerly accepted by his contemporaries, is idolized now winter power cuts, self sufficiency and conspicuous economy are in vogue.

6 Central Heating

Romans, with combustion gases circulating under the floor, were as far from central heating as Dr. Savot was with his fireplace that warmed an adjacent room. The insertion of vents allowing the heat of a hypocaust to rise to an upper storey, was the nearest the ancients came to a modern warm air system. True central heating is not achieved by merely concealing the furnace. It requires the introduction of an intermediary agent unrelated to combustion – air drawn from within or outside the building, or water and its co-efficient, steam – which is raised to a high temperature over a furnace fired by coal, gas, oil or electricity (called a boiler when the intermediary is water), and conveyed in pipes to radiators which, despite their name, transmit heat more by convection than radiation.

The earliest experiments in steam and hot water heating were made by horticulturalists. Hugh Platt in 1594 visualized in his *Garden of Eden* a hot-house warmed by steam piped from a tightly closed cooking vessel. The hot water pipes that he recommended for drying gun powder forecast those given by Sir Martin Trewald to his plants in 1716. Evelyn in 1664 planned for his flowers what is probably the first integrated system of heating and ventilation. His ideal conservatory had its vitiated air withdrawn in a suction flue to sustain an exterior furnace over which incoming fresh air was warmed in pipes. The air-tight room was protected by a vestibule admitting the gardener, but excluding draughts.

The cosy life enjoyed by plants was the envy of many humans shivering over sitting room fires. It gave the architect, J. J. Stevenson, the idea of heating the house and conservatory with one apparatus under the charge of the gardener, the only member of the household who could be relied upon to understand and respect its performance. It took an industrial revolution to bring the comforts of the conservatory to the home, to make, some might say, men into hothouse flowers. In 1784, forty years after Colonel William Cook published plans for steam heating an eight room house, James Watt had moderate success in applying this system to one room.

Necessity, not luxury, was the mother of central heating. New factories, offices, theatres, hospitals, large buildings occupied for long hours by many people, most of them sedentary, demanded new modes of heating that were efficient, clean, economical, time and space saving. The response to these demands was overwhelming, so too were the results. The first half of the nineteenth century abounded with uncertain but imaginative inventions jostling to convey heat by different means to groups of differing individuals at the right temperature and at the right place – up from the floor, down from ceilings, out of columns, under seats and windows.

Steam was pioneered by James Watt and Matthew Boulton in 1800 in a Manchester cotton factory, and widely publicized at Covent Garden Theatre where it was installed by the Marquis of Chabannes in 1816. Hot water circulated by a gravitational system was introduced in 1777 by M. Bonnemain to hatch chickens for the Paris market, and by 1832–3 was warming Westminster Hospital. The high pressure system, with smaller, more flexible pipes, perfected by Mr. Perkins, was more attractive to grand architects like Sir Robert Smirke who installed it in 1832 in the Print Room and Reading Room of the British Museum. Hot air pumped by a steam operated fan was forced down by Dr. David Reid on the heads of members in the House of Commons in 1834. A few years later, in the House of Representatives in Washington, D.C. it was forced up 'through an immense stack of dirty, rusty, iron pipes' into a 'labyrinth of uncleaned horizontal air-ducts' finally to emerge, at a temperature of 100 to 120°F, out of 'spitoons (originally intended for ducts) arranged all over the floor.'

Some of the less cumbrous contrivances were adapted for domestic use. The market, to begin with, was not large. Although there were a few adventurous or scientifically minded English customers, it was dominated, then as now, by Americans with new wealth, new homes to protect against an extreme climate, an abundance of fuel, and little or no attachment to open fires. Hot-air, being the cheapest and easiest system to install, was the most popular. But it was not very efficient without the aid of a fan which, before the 1880s when the electric motor was available, required a steam engine making a plant of prohibitive size. The heat, left to find its own way through vertical ducts, could only be used to supplement open fires or stoves. Ducts were usually introduced in halls or passages to warm the draughts, and create an overflow for adjoining rooms. After Lewis W. Leed's frequently quoted description of the ducts in the House of Representatives, hot air was widely condemned as unfit for human consumption. It certainly was not recommended in occupied rooms, least of all in class rooms for, worse than withering the brain and drying up talent, it was considered un-Christian, abusing God's gift of cool air to breathe and warm ground for the feet.

Steam and hot water circuits could provide greater heat to a larger area, but were more difficult and costly to install. Cost, however, could not cool the Duke of Wellington's love of heat. The hot water system put into Stratfield Saye in 1833 was among the first in private use. The blistering of paint on the walls suggests that it was also one of the hottest. Even in America in 1850, hot water heating, according to A. J. Downing was 'confined to town houses of the first class.' Confinement did not last long. In 1877 Birdsall Holly issued the residents of Lockport, New York with steam from the first district heating station. Within a few years steam began rising from the streets of New York City. The nuisance of boiler and fuel was removed from the home, so too was freedom of choice. But other nuisances remained. Pipes and radiators were eyesores, easily cured by those able to afford them in the first place. Pipes were concealed in decorative cylinders, or buried under floors, in skirtings, and later in walls to produce panel heating. For radiators that were not gilded or silvered making them radiate less heat than if they were blackened, there were perforated covers to suit all tastes. Those in the British Museum were complimented for looking 'exceedingly classical' in their marble cases. Ham House radiators were disguised and used as baroque side tables [*fig.24*]. The only incurable nuisance was the 'intermittent hiccough' of steam which still startles the uninitiated, but after prolonged encounter brought Proust memories of *temps perdu*, among them his first kiss to Albertine.

Undoubtedly, many of the early installations were imperfect, but the ill-effects for which they were censured were often due, as they still are, to the people and places they served. Central heating was blamed for poisoning the system, lowering resistance to colds, causing migraines, sapping talent, 'drying the life and substance out of about two-thirds of the people' of American along with their 'old pictures, furniture, and fine bindings.' It was also implicated in the deaths of hundreds of American business men, and of J. T. Smith, curator of the British Museum Print Room. J. J. Stevenson alarmed the public with a report that 'each year some English mansion with its accumulated wealth of art is burnt down by the overheating of a hot-air fire.' Instead of supervising their furnaces, many of his readers preferred to 'shut the hot-air inlet . . . and keep warm by putting on more clothing.'

The English, who freely chose to be half baked by open fires or scorched by the mid-day sun, were less opposed to overpowering heat than to the overpowering, totalitarian system which, unlike open fires, threatened to deprive them of full mastery over their personal needs, and was unescapable. Le Courbusier's dream of an absolute, universal indoor climate, whether its temperature was fixed at 18 or 10°C, would have been an Englishman's nightmare.

Overheating was a valid complaint against most of the

Figure 24. Radiator in the Round Gallery at Ham House, Surrey.

early systems, and a common one even from Americans before they became acclimatized. The automatic thermostat, pioneered in America in 1849, did not come into domestic use until the later decades of the century. Supervision of the furnace was left to ordinary household servants who had no instruction, nothing but coal dust to remedy the effects of overfeeding. Room temperature was extremely difficult to modulate with only a register over the hot-air duct or a stopcock alongside the radiator. In well built houses surfeited with dust-collecting furniture and drapery, with hot and smelly gas lights and no ventilation, the heat and odour was bound to be unpleasant. Centrally heated nineteenth-century sitting rooms, were they not separated from kitchens and water-closets, might have had as foul an atmosphere as sixteenth-century Russian stoves.

Open windows offered air free of charge, but they also admitted cold draughts which fell to the floor where least wanted, and let out expensive heat. Prejudice against unhealthy night air forbade the opening of windows when most needed. For the majority, the only alternative was a small hole in an exterior wall or chimney breast, fitted with one of a wide variety of patent ventilators: Dr. Arnott's with a hinged door, a metal grate plain or fancy, a straight forward air brick. Whether these were best placed near the floor or ceiling was debated along with the rise or fall of unidentified germs. The compromise was a Tobin's tube at middle height. Whatever their shape or place, the ability of these gadgets to evacuate a fug was at best minimal, nil when clogged with dust or draped, as they often were, to exclude draughts. Reliance upon them for fresh air was ridiculed by the pathologist, Dr. Ernest Jacobs, as no better than supplying 'a house with water by opening a trap door in the roof to admit rain.' The quest for higher standards of health and hygiene by late nineteenth-century medical men and sanitary engineers and by two remarkably liberated American women, Catherine Beecher and her sister Harriet Beecher Stowe (author of the anti-slavery novel, *Uncle Tom's Cabin*), made fresh air an object of cardinal, indeed moral importance. Public ills were blamed upon the incompetence of architects and builders, of men in general according to the feminist Beechers.

Only by co-ordinating heat and ventilation could fresh air be brought in without draughts, and stale air be extracted without loss of heat. The first comprehensive system was pioneered with difficulty in 1736 by Desaguliers for the House of Commons, London. Further development waited to be pressed by the needs of nineteenth-century hospitals, prisons, factories and other public buildings. Unlike central heating, large and speculative ventilating systems were rarely adapted for domestic use. Instead, outmoded fireplaces were put to work to extract the waste of mechanical heaters. At Osmaston, Derbyshire, built from 1846–9 by H. J. Stevens, vitiated air was drawn in by open fires and forced down into a duct under the kitchen garden where a giant stack waited to expel it. This was improved upon by George Davey who, in 1851, exhibited a scheme for drawing the waste of household chimneys down to sewers linked with a powerful furnace and finally with a colossal ventilating shaft. A small number of these 600 feet tall towers promised to replace thousands of chimney pots with flower pots. No one dreamed that a simple accident or failure along the line might reverse the system, until some boiler men at the U.S. Treasury, wanting to let off steam, opened the door of their furnace room onto the sewer.

With electricity came plenum systems using fans to introduce fresh hot air at a pressure high enough to force stale air out through gaps or specified vents. The air conditioning plant, blowing hot and cold, was invented by Willes Carrier in the 1920s, but did not appear on the market until after the last war.

The effect of central heating on twentieth-century life cannot be overestimated. Masonry walls are no longer required to keep us warm, neither are heavy curtains or interior doors. Owing to 'our effective hot water heating', wrote Frank Lloyd Wright in 1910, 'the forms of buildings may be completely articulated with light and air on several sides! ! and the compact, compartmented box of a house could now be spread . . . into a more organic expression.' Smog has disappeared from our cities, and Cinderella is now a liberated woman wearing minimal clothing all year round. Iced drinks are a treat in winter as in summer. Those who enjoy heated cars to transport them from heated rooms to hotter shops and offices rely on newscasters and windows to inform them of the seasons. Central heating has brough us as much freedom as the car, and has placed as heavy a burden upon our economy and natural resources. Strikes, shortages, Arab oil interests threaten to topple our mechanical idols, forcing us not to return to nature, but at least to look at the sun for warmth.

Bibliography

The only comprehensive histories of domestic heating are:

Bernan, Walter, *On the History and Art of Warming and Ventilating*. 2 vols., 1845.

Wright, Lawrence, *Home Fires Burning*, 1964.

The following list includes only those sources that have been used or quoted in the text. Page references are given for quotations.

Alberti, L. B., *Ten Books on Architecture*, tr. J. Leoni, ed. J. Rykwert, 1955. Bk.V, chap XVII, p.108.

Aloi, R., *Camini e Ambienti*, Milan, 1963. *Esempi Di Arredamento Moderno Di Tutto Mondo . . . Camini*, Milan, 1951.

Banham, R., *The Architecture of the Well Tempered Environment*, 1969.

Barran, F., *Der Offene Kamin*, Stuttgart, 1957

Beecher, C. E. and H. Beecher Stowe, *The American Woman's Home*, New York, 1869.

Boynton, L. ed., 'The Hardwick Hall Inventory of 1601', commentary by P. Thornton. *Furniture History*, VII, 1971. pp.10, 16, 31.

Boorde, Dr. A., *The Wisdom of Andrew Boorde* (*Dyetary of Helth*, 1542), ed. E. Poole, 1936, pp.22, 24.

Byron, Lord, *Letters and Journals*, ed. L. A. Marchand. Vol.III, 1974, p.226.

J.W.C., *Our Dwellings Warmed*, 1875, p.28.

Chambers, Sir W., *Treatise of Civil Architecture*, 1759.

Collins, P., *Changing Ideals in Modern Architecture 1750–1950*, 1965.

Cornforth, J. and Fowler, J., *English Decoration in the 18th Century*, 1974.

Crook, J. M., *The British Museum*, 1972, p.143.

Cross, A. ed., *Russia Under Western Eyes*, 1971.

Downing, A. J., *The Architecture of Country Houses*, New York, 1850, p.473.

Edis, R. W., *Decoration of Town Houses*, 1881.

Edwards, F. Jr., *Our Domestic Fire-Places*, new ed. 1870. *On the*

Ventilation of Dwelling Houses and the Utilization of Waste Heat from Open Fire-Places, 1868.

(Fiske, S.) An American, *English Photographs*, 1869, pp.192–198.

Frazer, Mrs. J. G., *First Aid to the Servantless*, 2nd ed. 1913, p.75.

Hardy, T., *Return of the Native*, Bk.II, pt.6, p.144.

Kelly, A., *The Book of English Fireplaces*, 1968.

Kroll, M., *Letters from Liselotte, Elizabeth Charlotte, Princess Palatine and Duchess of Orleans 'Madam'*, 1970

Leeds, L. W., *Treatise on Ventilation*, New York, 2nd ed. 1871, pp.26, 114 ff., 133.

Macquoid, P. and Edwards, C. H. R., *Dictionary of English Furniture*, 1954.

Meister, H., *Letters Written during a Residence in England*, 1799, p.163.

Memoirs of the Courts of Berlin, Dresden, Warsaw and Vienna in the Years 1777, 1778, and 1779, 3rd ed. 1806, p.262.

Montagu, Lady Mary Wortley, *Letters*, ed. R. Halsband. Vol.II, 1965, p.290.

O'Dea, W., *Making Fire*, HMSO, 1964.

Panton, J. C., *From Kitchen to Garret*, 1888, p.68.

Piranesi, G. B., *Diversi Maniere d'Adornare i Cammini*, Rome, 1769, ed. J. Wilton-Ely, 1972, p.7.

Proust, M., *Within A Budding Grove*, pp.296–297. *Sweet Cheat Gone*, p.108.

Richardson, C. J., *A Popular Treatise on the Warming and Ventilation of Buildings*, 1837.

Shuffrey, L. A., *The English Fireplace*, 1912.

Stevenson, J. J., *House Architecture*, 1880, II, chap.X.

Teale, L. P., *Economy of Coal in House Fires*, 1888.

Ware, I., *Complete Body of Architecture*, 1756.

West, T., *The Fireplace in the Home*, 1976.

Wharton, E. and Codman, O., Jr., *The Decoration of Houses*, New York, 1915.

Index

Printed in England for Her Majesty's Stationery Office by McCorquodale Printers Ltd, London
Dd 696346   C50